To Barbara and Ge[orge]
  Enjoy the secrets [...]
for you and your Granddaughters.
           Love
       Mable & Bill Cain
         2002

Beach Walks II

George Thatcher

[signature: George Thatcher]

 QUAIL RIDGE PRESS • Brandon, Mississippi

Copyright © 2000 by Quail Ridge Press

All Rights Reserved

No part of this book may be reproduced or utilized in any form or by any means, electronic or mechanical, including photocopying and recording, or by any information storage and retrieval system, without permission in writing from the publisher.

Cover photo by Donald Bradburn, M.D. • Printed in Spain by Bookprint, S.L., Barcelona

Quail Ridge Press
P. O. Box 123 • Brandon, MS 39043
1-800-343-1583

---

9 8 7 6 5 4 3 2 1

Library of Congress Cataloging-in-Publication Data
Thatcher, George 1922-
   Beach walks II / George Thatcher.
      p. cm.
   Includes bibliographical references.
   ISBN 0-893062-21-X
   1. Nature. 2. Seasons. 3. Beaches I. Title: Beach walks two. II. Title.
QH81 .T3392 2000
508—dc21

00-040278
CIP

To the Right Reverend Jackson Biggers,
Anglican Bishop of Northern Malawi,
East Africa,
who tills the Lord's garden
in a far land.

Author's note...

This book, a sequel to Beach Walks, is a new selection of my daily columns from the Sun Herald, a Knight Ridder newspaper, circulated on the Mississippi Gulf Coast.

Published two years ago, Beach Walks is now in its second printing, the first having sold out in a few weeks.

By-products of my daily walks on the beach, the columns began as journal notes. Each can be read in a minute or less, which suits newspaper readers' busy days. But it is hoped that readers of this book will enjoy only a few pages at one sitting.

Allow me to express gratitude to Gwen and Barney McKee, Cyndi Clark, and all the other great people at Quail Ridge Press, a superb independent publishing company. I thank the staff of the Sun Herald, especially Marie Harris, the editorial director, and Roland Weeks, publisher, for their countless efforts in transforming cryptic notes into illustrated, readable text.

The photograph of sanderlings on the front cover is the inspired creation of acclaimed nature photographer Dr. Donald Bradburn of New Orleans and Ocean Springs. To him I am very appreciative.

To readers of the first book, this sequel, and the newspaper column, I am humbly grateful for your kind reception.

George Thatcher
September, 2000

The Beach in
_____

Autumn

There is something curative in visiting the shore. Sitting on the seawall, one feels trouble blow away in the wind. Endless waves bring with them serenity. The sun warms the innermost chambers of the spirit. Once beset by discord, our Lord sought peace, ". . . Jesus went out of the house and sat by the sea . . ." Among others, Thoreau found its solitude, looking seaward, he wrote that he ". . . put all of America behind him . . ."

The Gospel According to St. Matthew 13:1. *Cape Cod* by Henry David Thoreau.

Tracks in wet sand tell a story this morning. First are those of a whelk, the small creature carrying its burdensome shell back into the sea. Then the prints of a large seagull. Both tracks meet. The drama ended a short time ago. All that remains now are an empty, upturned shell and a feather. Of course, calling the scene dramatic is exaggeration. It is merely one event among many occurring on the beach today.

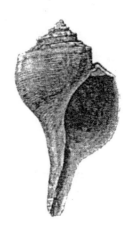

Pleasantly warm autumn days entice people to the beach. Today in the span of a mile, twenty-odd people of all ages enjoy the seashore. "Patrick loves Elizabeth" reads one message etched in sand. More cryptic, another declares "I" then a heart-shaped symbol followed by a well-drawn sheep which says, "I love (ewe)!" Farther along, a wondrous sand castle complete with turrets, moat, and drawbridge nears completion.

Envision, if you will, the primitive beauty of this coastland before the coming of European explorers and colonizers. The lowlands, cuestas, and coastal meadows were home to four tribes, the Biloxi, Pascagoula, Bayogoula, and Acolapissa, people who fished and farmed in a virtual Eden. They divided the year into seasons, lunar cycles, lyrically terming autumn, "Fall of the Leaf," giving other seasons similar names.

A History of Mississippi edited by Dr. R.A. McLemore, University Press of Mississippi, Jackson, 1973.

Seagulls, mostly laughing gulls, occupying a long sandbar this morning, are among the most banal of shorebirds. Some look for rarer species to add to their lists, but there is an august beauty in these gulls. So often we go far afield, seeking things which already lie within easy reach, the ordinary in our lives. "It's deadly commonplace," wrote Robert Louis Stevenson, "but . . . commonplaces are the great poetic truths."

"Weir of Hermiston" by Robert Louis Stevenson.

It baffles logic to consider the migratory flight of the little sanderling now busily feeding on the beach. Having migrated across immense distances, this fluff, an ounce or two of feather, flesh and bone risked all to come here. It stored body fats, made feather oils, and with its mate and chicks joined a flock, soared to great heights, found favorable winds. And after days and nights of constant flight landed here, an astonishing feat!

When first seen, the shell was discolored, yellowed, caked with dried mud. Now days later, cleaned and bleached by a hot sun, it is transformed into a thing of beauty, a pleasure to behold. Are there other things in life, people perhaps, repelling at first, but in reality attractive later? Butterflies were once hairy caterpillars. Blake said, "If the doors of perception were cleansed, everything would be seen as it is . . ."

"The Marriage of Heaven and Hell" by William Blake.

A sole gull, gliding westward toward a vanished sun, descends effortlessly on unmoving, extended wings. Its masterful flight is a thing of beauty. For a moment, the gull holds dominion in the sky, the only bird flying. The remainder of the flock is already at rest on the beach, quieter now as the blue sea pales into night. After a balletic landing, the gull joins the flock, hushed until morning's first light.

"*Evening*" is a beautiful word first crafted in Middle English language about the 12th century and used continually for 800 years. Once it meant that time from sunset to nightfall, but now generally from sunset to bedtime. The sun having set, it is evening on the beach now, a peaceful, quiet twilight. The evening star is hidden behind gossamer clouds. Long ago, this would be the hour for evensong, then the evening meal.

A piece of driftwood comes ashore this afternoon encrusted with hundreds of little acorn barnacles. It is said that they eat only when submerged in water. High and dry, their lives are now threatened. What strange creatures! With no eyes, no head, no gripping tentacles, a tiny "brain," the barnacle is a floating nomad, until it secretes a glue, cementing it to a host: a rock, a piling, concrete, a crab, or even this bit of floating wood.

The Ocean Almanac by Robert Hendrickson, Doubleday, New York, 1984.

A late rising moon, now waning, brings a high tide tonight. In autumn and winter, most high tides occur at night; while in late spring and summer, they occur during the day. As the Earth rotates tonight, our sea moves away from the moon, and the tide will move westward following the moon's path. In Japan there is more esteem for the moon than here. There people sometimes gather to watch its reflections on the water and to read poetry.

An acorn barnacle cemented to a piece of driftwood is sealed in protective armor, six hard calcium plates which defy attack. Rock-like, they are assembled aesthetically as if by a master jeweler. The barnacle's tiny brain tells it to feed an active digestive system, which consumes plankton, small animals, and detritus. Once Thoreau observed that Nature's total genius is directed toward perfecting its least important creatures.

Margins by Mary Parker Buckles, North Point Press, New York, 1997.

*E*ons ago sand on the beach was a part of the Appalachian Mountains. Now after some 60 million years, rock and earth, pulverized and ground, are grains of sand, fine as sugar, washed here through the ages. Geologists know the origin of the sand, because its mineral content is the same as the mountain range from which it came. Such is the power of Nature that it reduces mighty mountains to tiny, white bits of quartz.

Islands at the Edge of Time by Gunnar Hansen, Island Press, Covelo, California, 1993.

Compelling to watch, white caps break on the beach, ending in frothy, white foam. Every five seconds now, one pursues another. A talented poet, Charles G. Bell, watched waves, too. The waves are hasteless, he wrote, "What they are flows without effort into what they shall be. Their breaking is implicit . . . Then in a sudden splendor of foam, they boil ashore . . . finally breaking . . . rush back to the sea."

"Long Beach Island" by Charles G. Bell (1946), *Songs for a New America*, Indiana University Press, 1953.

The sea, a deep blue color yesterday, reflecting an even bluer sky, has acquired a beryl green tint mirroring today's gossamer clouds. Sailing the Mediterranean in 383 A.D., Saint Augustine remarked that the sea shifts colors "like veils green . . . or purple or heaven-reflecting blue . . ." Rimbaud too wrote, ". . . of the Sea, steeped in stars, milky, devouring the green azures . . ." What color will pigment tomorrow's sea?

The City of God by St. Augustine, Doubleday, New York, 1958. *Le Bâteau Ivre* by Arthur Rimbaud.

A little comb jellyfish lies on the beach a few inches away from lapping waves, its body still pulsing with life. Will it survive until the next tide lifts it back into the sea? So fragile, the jelly might not endure even the most caring rescue now. Intervene or not? For eons, sea creatures have braved similar dangers. No, Nature handles things very nicely on her own without meddling by a beachwalker.

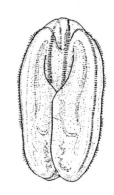

A whelk crawling at water's edge today is an example of common perfection. Armored in its protective shell, it lives out its life in oneness with nature, responding to rhythmic tides and suns and moons. "What beauty for contemplation," observed Saint Augustine, and what bounties . . . God has scattered like largess for man! What words can describe the myriad beauties of land and sea and sky?

The City of God by St. Augustine, Doubleday, New York, 1958.

An osprey has returned, soaring now 50 to 100 feet above the Sound on wide, powerful wings! Eyes focused downward, searching for quarry, the large bird commands the sky. As it flies above shorebirds roosting on sandbars, the osprey frightens them. Agitated and skittish, all fly away, leaving the raptor alone in its hunt. Soon the osprey dives into the water with a great splash, flying away to the north, its prey held in strong talons.

*O*nce, when Saint Francis was weeding his garden, he was asked, "Suppose you knew you were going to die later this very day, what would you do?" Without further thought Francis responded, "I would finish weeding my garden." What better thing to do? Walking the beach today amid the things and creatures of God's sublime creation, one ponders his own response to the question. The answer, of course, is "continue the walk."

$\mathcal{A}$ baby crab, barely an inch in size, crawls on the beach in dire danger of being eaten by nearby shorebirds. The crab, only a few drams of matter, quickened by that spark of Godhood called life, is keenly alert in its escape, darting excitedly on the wet sand. The instinct to survive is a primitive impulse in its flight to the sea. At last the little crab reaches safety, briskly vanishing into a whitecap breaking on the shore.

*T*he magical hour of bird migration is here! A small flock of plovers roosts well up on the beach, tightly gathered together, tired after their long southward flight. Unerringly, they have flown here at elevations of 2,500 feet from their summer nesting grounds. Rilke writes that migratory birds are "single-minded, unperplexed" unlike humans. And so they are. The birds' long journey calls to mind Plato's idea of a migrating mortal soul.

Duino Elegies by Rainer Maria Rilke. *Apology* by Plato.

A landsman, John Burroughs loved the sea and its shore, too. He noted that the smell of a sea breeze is tonic, that the beach is a larger, more primitive "out of doors" than inland landscapes. "This breeze . . . is a breath out of the morning of the world," a fountainhead, he wrote. The briny, pungently saline essence of our sea wind grows so commonplace that it takes a Burroughs to remind us that it is a benediction.

John Burroughs' America edited by Farida A. Wiley, Devin Adair Co., New York, 1951.

The complex relation of the sea to the shore and of all beach creatures to each other is cryptic. For example, the existence of the colony of jumping insects at water's edge today is not only a mystery, but a great miracle as well. Why did the Creator endow this low phylum with life? "My thoughts are not your thoughts," Isaiah quotes God; "My ways are not your ways . . . My ways are above your ways; My thoughts above your thoughts."

Isaiah, Chapter 55:8.

The beach is a natural landscape, one of sand, dunes, and a shoreline, but there is also another landscape here—a spiritual one. For example, this morning's serene sunrise calls to mind the hymn, "Morning has broken like the first morning . . ." of creation. A bird flies. Is it like the flight of that first bird? Is this sublime day akin to that primeval one? Is not this the same sun which rose over Eden?

The Hymnal 1982, No. 8, Church Publishing, Inc., New York.

Picture the setting. It is a foggy morning with no wind, the water placid and calm, not even a ripple. Near shore, a great blue heron stands motionless. The scene resembles the subject of a Japanese silk painting, or a haiku verse. The bird's reflection in the water, an undistorted mirrored image, brings it into even closer view. Unmoving heron, poised in hushed sea, waits, waits, waits.

For weeks now a large laughing gull, possibly in its third winter plumage, has been earth-bound with a broken or dislocated wing. Yet the gull not only survives, it thrives, fattening indeed, despite a crippling disability. To offset not flying, the bird has learned to walk rapidly. Now it roosts with a large flock, but when the flock flies away in alarm, the wounded gull alone runs toward the seawall.

*U*nusual sightings on today's walk: An ugly toadfish, sometimes called a mudfish. An oyster, along with many barnacles, encrusted on the back of a horseshoe crab. A large sheepshead, a fish probably tossed overboard by a passing shrimp boat. A very beautiful sharkeye shell, its occupant burrowing in wet sand. A family of four snowy egrets together, two adults and two immature birds, gaily gamboling near shore.

A moth flies among wilting wildflowers growing on the sand dune, pausing for a while at each blossom. Its narrow wings, brown and tan, speckled with black dots, flutter lightly. Perhaps the moth is saddened to find so little nectar in late-season blooms. Artist Walter Anderson once painted a moth, writing, "One hates waste and to have that life and that beauty" perish without notice "is becoming difficult to allow . . ."

The Horn Island Logs of Walter Inglis Anderson, Memphis State University Press, Memphis, 1973.

*T*he horseshoe crab laying eggs today is said to be unchanged in 350 million years. Other creatures may have evolved faster. Whether man once lived in the sea is not worrisome, if one views God as Lord of all things, including evolution. Paleontologist Pierre Teilhard de Chardin taught that nature produces ever more complex offspring, all converging at the end in the fulfillment of Creation, the Second Coming's ultimate unity.

The Divine Milieu by Pierre Teilhard de Chardin, Harper & Row, New York, 1957.

Until a sticker grabs a sock or pains a toe, the lowly sandbur escapes notice. Legion at seaside, it is an engaging (pun intended) plant. An annual grass favoring sand and dunes, its thorns protect against intruders. Spined burs contain seeds green at first then brown in late summer. Clinging to clothes, driven by wind, floating in high tides, seeds are scattered for propagation, stabilizing loose sand on beaches.

*O*n a warm redolent, pleasant November afternoon, a gentle breeze blows from the east. Overcast skies cloudy, but not threatening rain, radiate midday brightness. Offshore, an exposed sandbar holds a number of shorebirds, mostly gulls but also six brown pelicans, one white pelican, terns, skimmers, plovers and peeps. Wavelets lap at the waterline. All the beach is at peace, basking in the ambient glow of autumn.

Before sunrise there is quietude on the beach. Skimmers fish in darkness, their lower mandibles slicing calm water. Only muted highway sounds reach here. Then opalescent dawn comes, milky beach and sky, sand and clouds the same light gray color. Behind clouds, the sun rises on the eastern horizon, colorless, but tinting the western sky a pale pink soon dissolving into oyster-hued day.

Despite light pollution, minor meteor showers are visible this morning before daybreak. It is in fact a meteor shower rather than a storm. Consider the awe of early man eons ago standing on a beach seeing "shooting stars." Peering into the skies, he did not know that Leonid meteors, some as small as a grain of sand, rarely reach Earth. In the whole sweep of history, only six people and a dog may have been killed by meteors.

The New York Times, November 18, 1998.

A handful of sand may be a world in itself. Pining to study microbes in the next century, Edward O. Wilson, the famous scientist, guesses that there are 10 billion bacteria in a gram of soil. With a microscope, he would "cut his way through clonal forests sprawled across grains of sand," traveling through lake-sized water drops. All this adventure, he contends, lies within 10 paces of his laboratory. Wonderful things are close at hand.

Naturalist by Edward O. Wilson, Island Press, Shearwater Books, Washington, 1994.

$\mathcal{A}$lthough voices of many shorebirds are strident warnings to other birds in the flock, some have pleasing calls. Thoreau noted the snipe's "peculiar spirit-suggesting sound." The curlew's "sweet crystalline cry" was heard by Yeats and also by Dylan Thomas. In the 1600s Basho wrote about the "voices of plover." It is that plaintive, haunting whistle ("weeeeee") of the plover which is heard on the beach this morning.

The Wind Birds by Peter Matthiessen, Chapters Publishing Ltd., Shelburne, Vermont, 1994.

*S*ing a paean to the joy of flight, the ultimate freedom of moving through space, unhampered by earth's constraints. Shorebirds today fly, soar, glide, dive in blissful delight. Two royal terns flap their wings rapidly, stationary above the sea before headlong dives. Five brown pelicans, all in line, glide eastward with scant wing movement. Earth-bound, humans watch their joyful flight in wonderment.

The sun rose this morning at 6:08 and will set at 5:10. Man measures time in hours, days and centuries, but there are other clocks ticking, too. Time in cosmic space is marked in billions of light years, measurement beyond our ken. While our time is finite, celestial time is infinite. "Time," said Thoreau, "is but a stream I go a-fishing in." Our task is to escape the jail of ticking clocks and to enjoy the gladness of the moment.

Walden by Henry David Thoreau.

$\mathcal{A}$ sandy field near the small craft harbor is ablaze with the vivid colors of wildflowers this year. Formerly a shipyard, the area is overlooked except by chance visitors. Here bloom myriad flowers. But it is the blanket flower which is the most colorful with its deep red petals tipped in yellow. Jesus said, ". . . Learn from the way wildflowers grow . . . not even Solomon in all his splendor was clothed as one of them."

The Gospel According to St. Matthew 6:28-29.

There is a certain agreeable accord between shorebirds and regular beach walkers. A translation is, "Don't bother me, and I won't bother you," a sort of non-aggression pact. Seagulls, terns, plovers, and peeps happily abide by the treaty, while skimmers and herons nervously fly away when walkers approach. Chaperoning this morning, a friendly willet strolls only a few paces ahead, often turning its head to keep tabs on the advancing walkers.

*A*t sunset two gulls soar high in the sky, far higher than usual. It must be the pure joy of flight alone that propels them to those heights! Their normal sorties are hovering ones at low levels, foraging for food, but not now. In the twilight they soar and glide effortlessly, masters of aerodynamics. Their artful maneuvers should be recorded by a Saint-Exupery,* as they ascend and descend on wings of grace.

---

*Antoine de Saint-Exupery (1900-1944), French aviator and writer, author of *Wind, Sand, and Stars*.

A steady wind blows out of the east. Gusting to about 15 miles an hour, it is graded on the Beaufort Scale, a "moderate breeze," but strong enough to ground most shorebirds; even hardy pelicans float on the surface today. Offshore, the weather buoy at Chandeleur registers gusts to 20 knots, a "fresh breeze." Unlike Dickens, who felt uneasy in an east wind, people flock to the beach, braced by its refreshing force.

Several little fish have washed onto the beach overnight, least puffers, I think. They inflate themselves with water and air as a defense against predators. Lying dead this morning, they are still puffed and have sharp porcupine-like, small spines extending from their bodies. Unlike other dead fish, they are ignored by gulls and even by fish crows. Even in death their defense is effective.

Sometimes an unusual shorebird, perhaps a pair, appears on the beach during migration. Take for example the avocets, which were seen here a few days ago. Did they drop out of a large, migrating flock willingly? Weakened by exhaustion in a long flight? Were they blown off course by the storm? After a brief stay they departed maybe to rejoin the flock or to fend for themselves on a nearby beach. Yet the memory of their visit lingers.

During these days of hectic activity, one yearns for a simpler life, a little time of quiet peace, an escape from problem-filled days. The beach affords sanctuary, brief respite from life's reality, opportunity

to be alone, to empty one's mind of the strident calls of family and friends. Sit on the sand; feel the gentle breeze; listen to lapping waves. For a time forget nagging burdens. Savor the gladness of the fleeting moment. Leave refreshed.

Large numbers of jellyfish are beached this morning, deposited there by last night's high tide. Very fragile, the immature jellies (barely an inch in diameter) lie vulnerable, awaiting tonight's incoming tide. A problem with jellies is that they do not propel themselves and are moved by tide, wind, current and wave. An exception, "By-the-Wind Sailor," is a jelly able to tack like a sailboat.

Ranking beach creatures, Peter Matthiessen points out that there is more difference between jellyfish and eel than between eel and man. Shorebirds, he says, are more closely related to man than is the eel. High on the pyramid, birds "share not only our vertebrae, but our warm blood and central nervous system and not a few behavioral traits beside," he observes. Viewing a host of shorebirds today, one feels a closer kinship to them, perhaps cousins in the same phylum.

The Wind Birds by Peter Matthiesse, Chapters Publishing, Ltd., Shelburne, Vermont, 1994.

So distinctly marked in summer with jet black heads and wing tips, adult laughing gulls change in late summer and autumn. There is a marked graying as summer ends. Easily recognized a few weeks ago, laughing gulls have now become nondescript, and are often confused with other gulls. Still heard, however, in all seasons are the birds' strident, raucous calls, harsh laughs, "ha, ha, ha, ha, ha, ha, ha."

A small toadfish swims sluggishly near the beach in today's calm, high tide. Probably the ugliest fish in these waters, toadfish seen here are about 6 inches long. If, as the saying goes, beauty is in the eye of the beholder, is not ugliness also? Appalling in appearance the toadfish, along with us, is one of God's creatures. So we look past its gross features, knowing that humans to it may appear equally grotesque.

Walking is more mental and spiritual than merely taking physical exercise, wrote Henry David Thoreau. *It exceeds swinging dumbbells, counting strides, or timing the distance traveled. Walk like a camel, ruminating, he suggests, escaping present-day problems. "So we saunter . . . til . . . the sun shall shine more brightly . . . into our minds and hearts . . . lighting up our whole lives . . . warm, serene, and golden," he concluded.*

"Walking" by Henry David Thoreau from his journal, *Writing Nature*.

Trees on the barrier island, so clearly visible atop the seawall, disappear from sight as one descends to water's edge. The meager five-foot elevation difference shortens our ken. From the seawall, the sky far to the southwest was filled by the wooded island. Now at sea level, the horizon holds nothing at all except a line where sea and sky meet. The island, only nine miles away, dissolves into oblivion by the slight descent.

The mallow tree, growing beside the beach road, still has three lovely red, Confederate Rose blooms. The other blossoms have fallen victim to near-freezing temperatures and the soft rains of last week. One may think that late autumn is a time of demise and bereavement for nature, but it is actually a period of rebirth. A source of new life, the seed pods wither and burst, scattering particles of vitality which will become next year's flowers.

The Beach in

Winter

Gone is the warmth of last week's Indian Summer. By contrast, the beach today is cold, not frigid, but buffeted by a brisk north wind, unpleasant for long walks. During last night's rain the temperature was in the 70s, now the low 50s. Friends on the New England coast consider this temperature a heat wave, but Southern beaches are different. Walkers here are accustomed to warm days and gentle winds even in bleak winter.

At sunset today, a blue and pink halo extends across the southern horizon, blue at sea level and pink above. Reflected on the water are subdued tints of the sky's color, pink near the beach, blue farther away. Not the brilliance of others, say those in the Philippine Visayan Islands, our winter sunsets replace intensity of color with understated yet aesthetic tones which send artists to their canvases to capture fleeting beauty.

The dead laughing gull, lying inert at water's edge, probably met its demise yesterday. The bird is unmarred, showing no wound. Death is a frequent fact of life on the beach. And humans may err when they feel sorrow at the death of a beach creature. Nature, Muir observed, is always loving, healing all scars, all wounds, whether "in rocks or water or sky, or hearts." So we place the fallen gull in Nature's embrace.

John Muir In His Own Words edited by Peter Browning, Great West Books, Lafayette, California, 1988.

Ancient Romans celebrated this day, which they called "Invictus Sol," honoring the invincible sun. For us, it is the winter solstice, a day of brief light, when the sun is low on the southern horizon. Although winter looms, there is happiness in knowing that each coming day will be a bit longer, as the sun climbs higher in the sky. Overcast skies today hide the sun, as it begins its slow journey northward, bringing light and warmth.

A chilly 15-knot north wind makes today's beach walk cold. Early Greeks called the north wind Boreas. Then the Romans named it Aquilo. Mythical in origin, the source of all winds was once thought to be a cave on a Greek mountain. Homer's Ulysses held unfavorable winds in a bag. "The wind blows where it wills," Jesus said, "you can hear the sound it makes, but you do not know where it comes from or where it goes."

The Gospel According to St. John 3:8. *The Ocean Almanac* by Robert Hendrickson, Doubleday, New York, 1978.

The effect of last night's brief rain is visible on the sand beach this morning. Each raindrop left a slight depression where it struck the sand, marks that will remain for days. Strangely geometric in design, the patters are like pointillism in painting, that delicate technique begun by French impressionist Georges Seurat. Millions of rain marks embellish the sand this morning, adding new dimensions of beauty.

The wrack line, that stripe where refuse is left by the last high tide, is well up on the beach this morning. It comprises mostly grass, seaweed, leaves, tree bark, twigs and other vegetation. But, also there are shells, fish and crab remains, corks and lures, and numerous plastics, including hundreds of cellulose cigarette filters. Time, wind and tide will cleanse the beach, but Mother Nature does not know what to do with plastic.

Pierre Bonnard, the French impressionist painter, once observed, ". . . as for vision, I see things differently every day." So it is with all people. Things appear altered from day to day. Of course, it is not things which change, but only the way they are perceived. Trees gracing the shoreline, for example, are the same, but we see them in varying degrees of light. Too, our viewpoint differs. Sometimes we are joyful, sometimes somber.

The tolerance which shorebirds show one another is a continuing source of amazement. There they are, midway up on the beach, roosting close together: seagulls, terns, pelicans, godwits, snowy egrets, peeps, plovers. All share a common, quiet habitat. Most shorebirds are always well-mannered.

An exception, however, is the battle royal when seagulls feed. But now all biological differences are forgotten in a peaceable kingdom.

"The beach-lines linger and curl, as a . . . garment that clings to and follows the firm sweet limbs of a girl. Vanishing, swerving, evermore curving again into sight, softly the sandbeach wavers away" so Sidney Lanier, "The Poet of the Confederacy," described a beach. Was it the beach at Ship Island, where he may have been held prisoner? If so, the island's beach is preserved in verse, captured like a fossil in amber.

Quoted in *The Seas: The Poetry of the Earth* by Michelle Lovric, Courage Books, Philadelphia, London, 1996.

*S*t. Stephen's in Vienna, Notre Dame in Paris, St. Peter's in Rome, St. Paul's in London all are beautiful, stately, saintly cathedrals which delight the human spirit and excite the intellect. In a sense, is not the beach also a sort of cathedral, a place set aside which delights and excites, a place for meditation and prayer? A friend speaks jokingly of attending the church of Saint Horn Island. But there is truth in his words.

The sun, hanging low in the southern sky, will follow its course today, bringing 10 hours of daylight. Time, an elusive asset, can't be put in a bucket, or defined; a continuum of events, says a dictionary. Time can't be saved, so we know how valuable it is. An hour spent beachwalking is 4 percent of a day, 15 entire days each year. Is it time wasted? Or is it wisely spent, perhaps a path to tranquility?

A new millennium dawns, but things seem much the same. "Do not ignore this one fact," Saint Peter wrote, ". . . with the Lord a thousand years are like one day." Waves continue to roll ashore as they have for eons. The wind and the sea have a comforting permanence. According to Ecclesiastes, "a generation goes, and a generation comes, but the earth remains forever . . ." A sanderling feeds at water's edge just as it did yesterday.

2 Peter 3:8. Ecclesiastes 1:4.

"*This is the day which the Lord has made; we will rejoice and be glad in it,*" the psalmist intones. And so it is. A bright sun shines on a cobalt sea. Extending to the horizon, the sky, a deep blue, sharply contrasts the whiteness of the sand beach. Not only is this perfect day the Lord's, but all days cold, hot, wet, dry, even nondescript, are his. We live in his domain, and with the psalmist, our hearts sing, "*This is the Lord's doing, and it is marvelous in our eyes.*"

Psalm 118.

*O*ne day in the mid-1800s on an isolated island off New Hampshire, poet Celia Thaxter watched a lone sandpiper. "Across the lonely beach we flit," she wrote; "staunch friends are we." The bird, perhaps akin to sandpipers feeding on our beach today, is deathless now, preserved forever in her verse. Thaxter and her friends, important artists, writers, and musicians, formed America's first art colony there, finding the beach endless inspiration.

The Poems of Celia Thaxter.

It is the little things which delight, brightening the day. For example, the mild (3 mph) southeast breeze this morning warms cheeks which yesterday felt icy north winds. So it is in all the days of our life. A pleasant smile from a passerby, a kind word from a neighbor, an unexpected note in the mail, the touch of a hand, the aroma of hot tea— all are brief epiphanies of the day, mundane events, not momentous.

A palm tree, having floated ashore on yesterday's high tide, rests high on the beach. Drying now after months of floating in the Gulf, the tree is encrusted with hundreds of barnacles, affixed as if they were cemented to the trunk. The root system, still fully in place, comprises countless, tangled, string-like shoots which carried water to the tree's highest reaches. One wonders if this handsome palm once stood on an idyllic tropical island.

*C*ormorants are fewer in number this year than any other in memory. Usually arriving on the Coast in November, they depart in February. One day in 1962, two thousand were counted in the Sound, but this year they are scarce. Perching on offshore posts, they dive, and swimming underwater, find fish. However, this winter on their migration flight they may have found Delta catfish farms more alluring than our coastland.

Catalog of Mississippi Bird Records, Volume I by Gandy and Turcotte, State Wildlife Museum, Jackson, Mississippi, 1970.

"*There is a certain slant of light on winter afternoon . . . when it comes, the landscape listens, shadows hold their breath . . . ,*" wrote Emily Dickinson. And so it is this winter afternoon, the light deflected from a setting sun, partially cloaked by clouds. Elsewhere she observed that nature seldom uses the color yellow, "leaving all that for sunsets." And almost on cue the western sky momentarily takes on a saffron yellow hue.

Complete Poems of Emily Dickinson, Little, Brown & Co., Boston, 1960.

Two brown pelicans perch atop ruined wooden posts on Moses Pier a dozen feet above the water's surface. Heads pointed downward, the birds intently watch for small fish to swim by. Suddenly diving headlong with a great splash into the water, a pelican at once swallows its prey and sits again on the post, head down as before. Much the same now as Audubon observed them here in 1836, we watch pelicans with rapt fascination.

This is the time of year when diving ducks swim near shore. A flock of 14 lesser scaups paddles along just north of Moses Pier, diving from time to time, ignoring nearby fishermen. Earlier three goldeneyes had paddled by. Not good at flying, these ducks are expert divers, disappearing instantly, then surfacing sometimes 20 feet away. To reach flying speed they run on the water's surface with webbed feet before becoming airborne.

Daily tides ebb and flow with measured certainty, responding to moon and sun orbits. "Nature is nothing if not rhythmic, and its rhythms are many and varied . . . ," writes Ian Stewart. "Some are like a heartbeat; others like breathing," he explains. Newton's view of a clockwork universe is a comforting concept, one defended today by seeing sun, tide, even the shorebirds feeding all in cycle, all in natural rhythm.

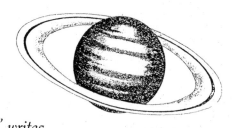

Nature's Numbers by Ian Stewart, Basic Books, A Division of Harper Collins, New York, 1995.

More jellyfish have washed up overnight, some larger combs than those found yesterday. Strictly not jellyfish at all, but ctenophores, they are "vagabonds" of the sea, as one writer calls them. Young combs grow better when water is saltier; they are nurtured this year by lack of rain. So fragile and so gelatin-like, yet so filled with that spark of godhood, life, that this one, when mature, may produce 40,000 eggs in a day!

National Aquarium Website, Baltimore, Maryland.

Today on remote Tierra del Fuego beaches, flocks of plovers prepare for the annual spring migration coming perhaps over our shores, as they fly northward to nesting areas in the Arctic tundra. Their ghostly passage follows watercourses, some believe, which existed before the Ice Age. Many plovers spend the winter here, old friends, observed by beach walkers daily, distant cousins of those in the South American flocks.

*T*wo very large flocks of black skimmers rest on the beach this afternoon, but frightened by a dog, a flock flies away. Their flight is spectacular! Turning, darting out over the water, the flight of nearly 60 birds is close-knit; flying as if they were but one bird, they land in unity farther down the beach. Now the dog reaches the other skimmers, who fly away, but in disarray, landing in small groups, barking in complaint.

At sunset on a hot day in Africa in 1949, Albert Schweitzer was struck with the necessity of revering all living things, affirming life as a spiritual act. Viewing the beach today—birds, crabs, fish, even insects at water's edge—one knows that he shares a gift of life with all creatures, that spark of godhood. Receiving the Nobel Prize, Schweitzer said, "You don't live in a world all alone. Your brothers are here too."

The Animal World of Albert Sweitzer, The Ecco Press, Hopewell, New Jersey, 1950.

*O*n an otherwise dark gray day, there is a narrow, luminous strip of peach-colored light along the southern horizon. In a strange way, the lighted horizon gladdens a bleak day, reminding us that it is not the big things which bring happiness, but rather small, insignificant events. The lighted horizon tells us that somewhere out there the sun shines through dark clouds. It is a small epiphany, like receiving an unexpected smile.

An ebbed tide this morning has left tidal pools of sea water. From mounds, worm-like animals protrude their heads, feeding (one guesses) in the calm water. Gathering such specimens once, John Steinbeck suggested looking up from a tidal pool to the sky and then back again, which puts all things in perspective. The sea worms seen today are as much a part of this wondrous universe as are the stars and people like you and me.

An Island Out of Time by Tom Horton, W.W. Norton & Co., New York, 1996.

The beach changes with the mood of the walker. On different days, one sees things differently. William Cullen Bryant wrote about the change. Nature, he noted, "speaks a various language; for his gayer hours, she has a voice of gladness and eloquence of beauty, and she glides into his darker musings with a mild and healing sympathy." It is true, of course, but puzzling, how the beach meets different needs from day to day.

"Thanatopsis" by William Cullen Bryant, *The New Oxford Book of American Verse*, Oxford University Press, New York, 1976.

*O*ften we overlook treasure underfoot. This very moment in time is all that we really have. Usually we are chained to both past and future. As T.S. Eliot noted, "All time is unredeemable . . ." Time past and time future coincide now. So it is that one stops to sense the fresh breeze, see the waves, watch the birds . . .
to live this one hour with no regard for what
has been or what may be.
This very moment is precious
treasure!

"Burnt Norton," *T. S. Eliot: The Complete Poems and Plays*, Harcourt, Brace & World, Inc., New York, 1971.

A lone golden aster still blooms on the sand dune. All other flowers have vanished, perished perhaps in last fall's rain and cold winds. Supposed to flower from July through October, this small yellow blossom has survived nearly three months beyond probability. Profuse in early autumn, countless florets covered the dune. Now all are gone except one. But they are not dead; their seeds will burgeon again in summer.

A fossil, held in hand, is the impression of a small fish on hard rock. About 40 million years ago, it swam in what is now the western part of the United States. It died and sank into a deep, soft sediment which hardened into stone, preserving the fish's skeleton. So it is that we hold at this moment a being that lived in the Eocene epoch of the Cenozoic era 40,000 millennia ago.

It is easier for seagulls to survive in summer than in winter. In summer the sea brims with life—fish, crabs, shrimp, plankton and jellies—some of which end up on the beach as food for carrion-eating birds. However, in winter marine life retreats into deep water, and the shoreline is often barren. Today, a large flock of hungry laughing gulls fights noisily over a small apple core floating in shallow water.

*S*ometimes during the course of a beach walk, one feels attuned to the harmony of his surroundings. Difficult to explain; it is a joining somehow of sea, wind, sun into a wholeness. "The sun shines not on us, but in us," wrote John Muir; "the rivers flow not past, but through us." Today there is that merging of the elements. One feels sun and wind, not separately, but together with a perception of oneness—nature's concord.

John Muir In His Own Words, edited by Peter Browning, Great West Books, 1988.

This warm January afternoon is ideal for sailing. Yet only one boat, a sloop, is seen across the entire horizon. It scuds along close-hauled, pointing up into a south wind. Unused, scores of sailboats lie at their moorings in the harbor, gently rising and falling with the wavelets, lines sounding against metal masts. Straining at their tethers, boats were made for sailing, not anchorage. Where are the sailors this fine day?

At intervals along the beach, black silt accumulates. Under a magnifying glass, the silt is seen to be bits of leaves, bark, twigs, pine straw, grass and sometimes seaweed all crushed by tide, wave and wind. A handful also contains grains of crystalline wet sand, clinging to the vegetation. Perhaps the now-fragmented leaves and other debris once grew on a barrier island or maybe on the mainland, before blowing seaward.

The sea, lying before us today, so blue, so pacific, is called the "great mother of life" by Rachel Carson. Out of its fertility is not only the edible, but other remarkable life forms—plankton, fungi, bacteria, and plants to name only a few. A bountiful pasture, the sea flowers, bringing forth countless species. Unknown to science, it is said, are perhaps 99 percent of all things living today, let alone those lost during the past three billion years.

The Sea Around Us by Rachel Carson, Golden Press, New York, 1958.

The heavy, chalky clam shell found on the beach this afternoon is large (5 1/2 inches). More common are smaller sizes. Scarred by surf and sand, the shell retains a pristine beauty in its delicate, concentric growth lines. Since earliest times, seashells have captured man's interest. Greek and Roman pottery was adorned with shell paintings, as was Renaissance art. Even today shells are popular items in souvenir shops.

A waning moon partially lights the beach tonight. Diminishing now, it will be replaced by a new moon in 10 days, meaning that February has no full moon. It was this celestial body of which Juliet complained in calling it "inconstant." If we feel cheated of a full moon this month we should remember that in January there were two full moons, as there will be in March. The 29-day moon cycle does not quite fit into February.

For a long time today, a brown pelican floated in the sea near the beach. Swimming slowly, bill piercing the surface, the pelican repeatedly dipped its head in the water, coming up each time with a catch. Then, lifting its bill, it swallowed.

So good was the fishing that the pelican, usually wary of humans, was oblivious to nearby beachwalkers. After an hour of feeding, the pelican flew to an offshore post to bask in the winter sun.

*O*bserved yesterday at seaside, the minute insects swarm once again today. Even under magnification, it is difficult to see them in detail; yet one knows that they (like all insects) have six legs, eyes, sense organs, and a brain. This colony of gnat-sized creatures lives in its own universe, not only beyond our knowing but perhaps our interest. These insects and 800,000 other species live their life spans heedless of us and our world.

These are the days when birds begin their long-distance migrating flights, sometimes resting briefly on these shores, sometimes continuing a ghostly passage to northern havens. Today some birds that we see—cormorants, plovers, turnstones and others—will leave for a distant Canadian tundra. But others, thankfully, will remain here, permanent residents. Soon the first least tern will appear, having wintered perhaps as far away as Peru.

$\mathcal{V}$ery early this year, little clover plants grow next to the beach road. Normally seen in April, the plant, abetted by unseasonably warm weather, produces ten minute yellow stamens before the leaves are fully developed. Nature lays before us a golden carpet of miniature blooms from plants which have escaped the threshing of the mowing machine. What beauty the world displays in unexpected places, if only we would look!

$\mathcal{A}$ drab winter day brings somber grayness to the beach . . . gray sky, gray sand, gray sea. The world appears devoid of color. In dull gray winter plumage, shorebirds roost on the beach. Yet near the end of the walk on this ashen day, two snowy egrets bring a bit of color and gladness. With golden feet and gold at base of bills, brilliant white wings and bodies, they brighten the day, frolicking in shallow water.

After an absence of three months, the family of fish crows has returned, looking a bit larger now than in the fall. Where did they spend the winter? Perhaps on some warmer Florida beach? Perhaps at an exclusive spa there? However, their return is not celebrated by nearby seagulls. Vying for carrion on the beach today, crows shoulder gulls out of the way, showing their dominance, flaunting their strength.

About 1855 Walt Whitman, usually upbeat and confident, walked a beach despondently. In desolation he wrote, "I too but signify at the utmost a little wash'd-up drift, a few sands and dead leaves gather . . . and merge myself as part of the sands and drift." One cannot sense the depth of his depression today, a glorious day filled with sunshine, wind, waves, contented shorebirds— all under a cloudless sky.

"As I Ebb'd with the Ocean of Life" by Walt Whitman, *Leaves of Grass*.

Today a squally wind blows away mental cobwebs, cleansing the mind of all clatter, emptying brain cells of the day's tensions. All anxieties brushed aside, the walk is a purely peaceful experience. An empty mind simply feels the brisk southeast wind, hears waves coming ashore, watches birds in flight. Time, for a bit, is suspended. With no concern for tomorrow or yesterday, the reality of this moment alone is enjoyed.

One would hardly notice them, because they are above usual fields of vision. But there they sit, three of them, mature great blue herons perching atop electric poles on Moses Pier! Seemingly unconcerned with the fishermen below, they stoically face into a mild north breeze. Normally herons, easily spooked by humans, fly away at their approach, but not today. Among the fishermen, only one stops casting to view the birds.

For a long time today watched several balletic willets feeding in the calm shallows of an incoming tide. In deeper water than usual, the birds continually submerge their entire heads well below the surface, finding small marine creatures. Now belly-deep, they are a drab gray until they fly away, showing flashy black and white wings and a white rump. The aesthetic, graceful, busy willet is everyone's favorite shorebird.

The sun rises this morning, a huge red ball, still partially below the eastern horizon. Soon, fully visible, it illuminates the entire sky with its brilliance. How fitting it is that today is The Epiphany, the beginning of that season of light sandwiched between Christmas and Lent in the church calendar. But there are small epiphanies too, those found in a stranger's smile, sudden rainfall, even the arrival of an unexpected letter.

We live in the midst of mathematical patterns, writes Ian Stewart in his book, Nature's Numbers. Stars move in circular orbits; intricate waves roll the sea, as does sand on this beach; the number of petals on a flower are in arithmetic sequence; zebras have stripes; leopards, spots. Nature's laws have to do with numbers. A great mathematician, the lord of the universe, governs not only the big things, but every atom, every molecule.

Nature's Numbers by Ian Stewart, Basic Books, A Division of Harper Collins, New York, 1995.

The Beach in _____ Spring

An hour after sunset with the last vestige of daylight gone, night comes. Beach creatures fall silent. Known more for art than for his poetry, Michelangelo described day's end, "O dark yet lovely night, the gentlest time, you seal with peace the work of every hand and only those who praise you understand."
A young moon, the equinox moon, adds scant light. And as winter ends, there is a seal of peace tonight.
Spring begins.

Michelangelo by George Bull, St. Martin's Press, New York, 1995.

Yesterday and for the last few days, gale-driven waves have rolled ashore. Windy weather caused small craft warnings to be hoisted. Yet today the sea is a placid calm. Changeful, today's sea may bear little resemblance to tomorrow's sea. For those who live near it and for those who venture upon it, Herman Melville advised, "When beholding the beauty of the ocean skin, one forgets the tiger heart that pants beneath it."

Moby Dick by Herman Melville.

A mullet jumps high in the air, again and again—three mighty leaps! Some say the jump is to avoid a predator or to find oxygen. But could it not be simply for the joy of flight, escaping a fish's natural limits, moving for a moment into a different sphere? Consider the mullet's mighty will to jump, its muscular energy, powerful tail movement, finally breaking the surface, then soaring through space repeatedly—a noble effort!

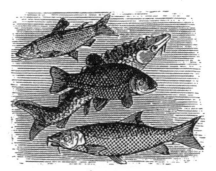

A seagull, flying above the beach, does not flap its wings, but descends with no body movement, gliding into a mild wind. Masters of flight, gulls seem to defy the rules of nature that govern airplanes and other birds, finding small thermals, sometimes soaring to new heights without the flap of a wing. Now the gull lands effortlessly, wings still outstretched until the movement of cushioned impact. No ballerina shows more grace.

This morning pelicans fly and dive a short distance offshore. Once the bird was an important religious symbol. Mentioned in a psalm, its likeness adorned crucifixes, paintings and cathedrals throughout Europe in the Middle Ages. Referring to Christ as "our own pelican," Dante lifted the species to lofty heights. In modern times, the pelican's holy image is forgotten, but the bird is still admired by one and all.

"Christ Our Pelican" by Ann Rose, *The Living Church Magazine*, May 2, 1999.

"Is no species other than man bored?" asks Walker Percy, pointing out that the word did not even appear until the 18th century. Too much happens at the beach to experience boredom here. Changing almost by the minute, the seascape is always in a state of flux, always presenting itself in new garb, always demanding attention. Bridie writes that boredom is caused only by satisfied ignorance or feeble attention levels.

Lost in the Cosmos by Walker Percy, Simon & Schuster, New York, 1983. Mr. Bolfry by James Bridie, 1943.

"*Love all of God's creation . . . every grain of sand of it . . .*," adjures Dostoyevsky. His advice is easy to follow on such a day as this one. The world blooms, and the beach is filled with nesting birds. Spring brings very special beauty to us. Edna St. Vincent Millay responded once, "*World, I cannot hold thee close enough! . . . Lord, I do fear Thou'st made the world too beautiful this year . . .*"

The Brothers Karamazov by Fyodor Dostoyevsky. "God's World" by Edna St. Vincent Millay.

*T*hree beautiful shiny whelks leave tracks at water's edge today, moving into gently lapping waves as they crawl toward deeper water. They lay long strings (some two feet long) of egg capsules, anchored in shallow water by the female until the eggs hatch. Afterward, egg cases are sometimes washed up on the beach, attractive enough to be worn as necklaces. Underwater now, the snail tracks are still visible until erased by waves.

Several conchs lie at water's edge today, burrowing in wet sand. When picked up, the conch's body retreats audibly into the innermost recess of its shell, seeking protection from a predator. The movement of its retreat is felt through the spiral chamber. What dread the conch must experience, awaiting attack from an unknown enemy! What relief it feels now as it is returned gently to the beach, safe and unharmed!

"May you live all the days of your life," was Jonathan Swift's wish for his friends. Scholars assume that he meant to truly live, to be fully aware, to savor every moment of existence. Also on the subject, Thoreau worried about people living in "quiet desperation." Walking the beach, one has the choice to live thankfully this day, alert to all its endowed beauty.

Given so freely, this day and those to come are treasures beyond measure.

Polite Conversation by Jonathan Swift. *Walden* by Henry David Thoreau.

At water's edge this morning are several orange-gray colored sponges—pieces, one guesses, of larger sponges appended to offshore oyster beds. Primitive animals, sponges were once thought not to be living creatures at all. The sponge from which the pieces were torn will soon replace these lost cells. Showing little change in 250 million years, fossils affirm that sponges were among Earth's earliest living things.

More horseshoe crabs on the beach today; four overturned, but when righted they crawl back into deep water. A story is told of a little girl on a Cape Cod beach, turning dozens of stranded horseshoe crabs, enabling them to return to the sea. Finally, her father said, "There are too many of them, honey; let's stop. We can't make a difference here." "Let me do one more, Daddy," she replied. "It will make a difference to him."

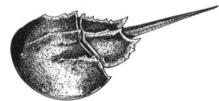

*I*n Arabic, the word desert is translated "sea without water." Looking at the sand beach today, one understands the comparison. Writers have spoken of "ripples of sand" and "waves of dunes." They are visible in the sand now, the results of wind waves coursing over the beach, designs similar to marks left by waves on the sea bottom. There is a moiré effect, artistic repetitive patterns, both on the beach and on the sand flats.

Man is not born only at the moment his mother gives him birth, but again and again at other times as well. There are new beginnings, moments of discovery and insight. Watching least terns nest today is an example. The whole flow of life on earth is imparted as minute birds defend nests, eggs, mates. It is as if all the forces of creation are met here at this hour, a revelation of life as it is shared by all living creatures.

A nearly full moon hangs over the beach, a familiar sight to all generations from the beginning of time. To early man, the moonscape was one of seas and mountains. Fanciful names were given, such as "The Sea of Vapors, The Ocean of Storms, The Sea of Serenity," all in Latin. Later astronomers found not ocean, but bone-dry rock. Yet the romantic names endure. And tonight "The Sea of Tranquility" is visible through wispy clouds.

A reality of time is that all we have in this present moment. "All time is unredeemable," says T.S. Eliot. The challenge is to be fully alive with all senses alert to the happenings of this one day. For me the reality of today is a mullet jumping, a wildflower blooming by the seawall, the cry of a plover, a friend's visit, a chess game, a cup of jasmine tea—all banal perhaps, but each has blessed, in its own way, the joy of today.

"Four Quartets" by T. S. Elliot, Harcourt, Brace & World, New York, 1952.

$\mathcal{A}$ mullet fisherman, armed with a throw net, wades into the sea this morning. It is a scene which has been enacted for thousands of years, back to the dawn of history, man pursuing fish. First for food and then for sport, fishing still occupies a place in the human mind. Although a huge gulf of time separates the wading man today from an ancient Egyptian fisherman, they both share the same gear and the same expectancy.

There are many interesting books on beach lore. All are fun to read, but nothing is better than finding out things on your own. Even the most gifted writer cannot describe fully the elegant flight of seagulls now overhead. Telling of their grace and beauty falls short somehow, missing their lissome, aesthetic passage through space. Once seen, however their flight becomes keen memory, ready for instant recall.

Some oceanographers believe that early in the planet's life the oceans contained fresh water, becoming salty later from rains and rivers. Changing salt content nearer land threatens marine life. Here on the Coast salinity varies after heavy rains. Found on the beach recently are many oyster shells. Once an important industry here, oysters are sensitive, not only to pollution, but also to changing salt levels in the sea.

*O*ne stoops to pluck a small blue flower, growing among the weeds and brambles by the beach road. But wait. Pick a flower? What a waste! Even in a watered vase, it would survive only a day, then lose its beauty. Was it Francis Thompson, the English poet, who said that plucking a bloom troubled a star? And Blake who saw heaven in a wildflower? The little flower remains untouched, perhaps to be seen by someone else.

The vastness of the universe boggles the mind! For example, starlight reaching the beach tonight, one reads, began its descent 30,000 years ago! "Beyond plants are animals, beyond animals is man, beyond man is the universe," writes a poet. And here we stand tonight, peering into the heavens starstruck, much the same as a Paleozoic man. Still beyond our ken now, the universe, once fathomed, will reveal the mind of God.

"The Blue Meridian" by Jean Toomer, Harlem Renaissance poet.

A monarch butterfly flutters northerly in heavy traffic on the beach highway, narrowly missed by fast cars. As if it led a charmed life, the butterfly, tossed wildly by traffic air currents, survives auto after auto. Crossing the eastbound lane safely, it now flits bravely into the westbound lane. Amid even heavier traffic, the painted-winged creature somehow reaches sanctuary, continuing its spring migration.

The beach in winter becomes the beach in spring with the passing of the vernal equinox only days ago. "Nothing is so beautiful as spring," writes Gerard Manley Hopkins. It is that peerless season sandwiched between winter and hot summer, when beach creatures and plants take on renewed vitality, a time of sunshine and breezy winds, a time of kite-flying, and of longer walks. "All Nature seems at work," Coleridge observed.

"Spring" by Gerard Manley Hopkins, *The New Oxford Book of English Verse*, Oxford University Press, London.
"Work Without Hope" by Samuel Taylor Coleridge.

Artists find the sea a willing and cooperative subject to paint, but finding precisely the right light and color is a problem. Complaining, Monet said that light disappears, lasting sometimes three or four minutes. To paint the sea, he explained, "You have to see it every day, at every hour and in the same place to come to know the life in this location . . ." On the beach today, an artist paints and repaints a canvas, trying and rejecting each tone of blue.

The Artist Speaks: Monet, edited by Genevieve Morgan, Collins Publishers, San Francisco, 1996.

A small piece of driftwood washes ashore this morning, riddled with wormholes. Dwelling deep within the cavities are single barnacles along with little clams. Green algae, gracefully moving in the wavelets, trails in the water. A world in itself, the driftwood is host, not only to life forms seen with the eye, but also to thousands of minute creatures thriving in life cycles beyond our comprehension.

A laughing gull lies dead this morning at water's edge, prompting a fellow beach walker to remark, "Well, there are so many; we'll never miss one." But Maimonides to the contrary said, "In the realm of nature there is nothing purposeless, trivial, or unnecessary." The gull's death diminishes us. Is it important? Surely there will be no newspaper headlines, but the God who knows every sparrow's flight surely records our fallen gull.

The Guide for the Perplexed by Moses Maimonides.

The majestic sea, stretched out before us today, was given a magical name, "oceanus," by the ancient Greeks. In the beginning they thought it to be a mighty river flowing around a flat earth. Shrouded in mystery, fable, and fantasy, the sea, despite millennia of study, still beckons our scrutiny. The seas, Rachel Carson wrote, all return "to Oceanus . . . like the ever-flowing stream of time, the beginning and the end."

The Sea Around Us by Rachel Carson, Golden Press, New York, 1958.

Who has not held a seashell to his ear to hear the roar of a distant ocean? Of course, you say it is not the actual sound of crashing waves, but merely an audible echo, a noisy din. Wordsworth termed it the shell's "mysterious union with its native sea." There are times, he said, when the shell relates "authentic tidings of invisible things; of ebb and flow, and . . . central peace."

"The Excursion" by William Wordsworth, 1814.

Four years before his death in 1862, Henry David Thoreau observed such a morning as this one. "These are true mornings of creation, original and poetic days," he recorded in his journal, ". . . such a mist as might have adorned the first morning." Obscured at daybreak by a gossamer mist, the distant horizon is now revealed in defined clarity. Does not each ensuing dawn remind us of creation's first sublime day?

Journal of Henry David Thoreau, 1858.

Yesterday's high tide left a line of brown and white pebbles. Gently lapping waves reach for them today. Under a magnifying glass, they are not only little stones, but also bits of crushed shells, fish bones, and tiny bivalves—some still intact. Lying together in a straight line, the white pebbles are sand-colored, rubbed smooth by tumbling in surf. Marked in shades of brown, the others are hardened clay with strata veins visible.

𝓕inding food is a continuing struggle for marine creatures. At water's edge this afternoon is a horseshoe crab encrusted by a dozen or so barnacles with conchs dining on the barnacles, an example of the food chain in action. The host entertains both the eaters and the eaten, while it devours clams. And large fish eat small fish. Food is life's first necessity, and the sea provides it bountifully for all creatures.

*K*eeping one's eyes open, simply looking at things on the beach, is far from a waste of time. Finding a shell, gazing at waves, seeing a conch crawl, detecting a jellyfish in distress, watching birds fly—all spark a feeling of wonderment. One observation leads to another and eventually to a lively interest in the beach and its creatures. Beholding shore life is exploring nature's frontier, that magical meeting place of land and sea.

*T*he joy of flight! All afternoon a flock of royal terns flies in search of food. Some 15 feet above the water's surface, a tern spots a small fish. Beating its wings rapidly to remain aloft but stationary, the bird suddenly dives headlong into the sea, airborne now once again with a minnow in its bill. Replete, the flock of six large terns and four smaller ones rests on a submerged sandbar despite the rising tide.

The vivid blue of the sea today, ultramarine, all but defies description, so intense is its color. The Latin root word, "ultramarinus," means beyond the sea, but today we use it to define a deep color of blue. Reflecting an azure sky, the sea is calm except for mild swells, but its blueness is compelling, the color, Ruskin once said, "everlastingly appointed by the Deity to be a source of delight."

Lectures on Architecture and Painting by John Ruskin, Garland Publishing, New York, 1978.

A new month arrives tomorrow on the heels of another fleeting month! "The months and days are the travelers of eternity," noted Basho, and "The years that come and go are also voyagers." Some 300 years ago, the Japanese poet watched a solitary cloud drifting in the wind, similar to the lone cloud today floating from the southeast horizon over the beach into the northwest sky. To Basho such clouds herald the passage of time.

"The Narrow Road of Oku" by Matsuo Basho, 1644-1694.

Most cormorants left this beach weeks ago, but one remains. Daily perched on offshore posts, spreading its wide wings, diving for fish, the bird is seemingly healthy. Why did it not join other cormorants in their migrating flight northward? This afternoon, quite near the beach, it swims in very shallow water, head submerged for minutes at a time, occasionally lifting its bill to swallow small fish.

The beauty of ordinary things—grains of sand, a bit of driftwood, the cry of birds, a humid breeze, lapping waves—the mundane escapes notice much of the time. Amid our hectic activities, there is scarce time to consider the commonplace. St. Thomas, in his apocryphal gospel wrote, ". . . lift the stone and one finds the Lord . . ." Divinity, enshrined in simple things, awaits those who seek, who listen, who look.

The Gospel According to Thomas, translated by A. Guillaumont, Harper & Row, New York, 1959.

"*The soul has an absolute, unforgiving need for regular excursions into enchantment,*" writes Thomas Moore; "*it needs them like the body needs food, and the mind needs thought.*" *In a modern world which crowds out spiritual enchantment, a solitary beach walk still provides opportunity. The act of being alone, watching elemental forces—the tides, the wind, sun, moon, and stars—brings to everyday life the mystery of our existence.*

Re-enchantment of Everyday Life by Thomas Moore.

Sitting on the seawall, looking at the sea, one wonders how it was long before Europeans came. Once, of course, it extended far to the north. A neighbor has a petrified oyster (50,000 years old) which he found 300 miles inland. But later the gulf ended here, and men came across the Bering land bridge to live at this place. Here they planted crops, fished, gathered shells, and perhaps admired the sea, much the same as we do today.

*A*nother sun sets in a cloudless sky. After dropping from sight, it brightens the horizons to the south and the east with a band of low-level color not found in the Munsell color ranges, a bright pink.

Reflecting the sky's pigment, the water takes on a similar hue, tinting the sea with a roseate glow. Not the ocean described by Patrick O'Brian as "wine-dark," the sea tonight is dyed a pink, one in which "sailors delight."*

The Wine Dark Sea by Patrick O'Brian, W.W. Norton & Co., New York, 1993. *"Pink at night, sailors' delight; pink in the morning, sailors' warning." —Old saying.

The beach, that microcosm of Creation, exhibits a wide array of Nature's wonders—sea, tides, winds, fish, birds, insects, sand—all within range of our sight, touch, hearing, and smell. Saint Columba, a 6th-century Irish monk, wrote that if one wished to know the Creator, then one must first understand His creatures and all His creation. Contending that Nature was a key to unlock divine mystery, he urged its study.

A Mapmaker's Dream by James Cowan, Shambhala, Boston, 1996.

One of the most popular postcards in the 1930s was a photograph of a dozen or so brown pelicans perched on every piling of a ruined pier. Tourists and others bought thousands of the card. Now some 60 years later, the scene is re-enacted! Eleven pelicans occupy posts of another pier, quietly facing into the wind, bills pointed downward, eyes focused on the water below, waiting for a target of opportunity to come into view.

Ten seagulls have occupied an exposed sandbar for hours this morning. Not moving, they stolidly face into the mild southeast wind, silently watching a calm sea. The tide rises, covering the sandbar with an inch or two of water. Yet the gulls remain impassive with wet feet. When the tide completely floods the sandbar, all ten gulls become waterborne, paddling slowly into the wind rising and falling with each gentle wave.

A colony of black ants living in a sand dune has two columns of workers, moving in opposite directions along a distance of 60 inches. One column carries bits of food—the remains of a dead grasshopper—into an underground nest. The other column returns to the prey for yet another load. Not much understood by humans, ants live in a domain all their own, a complex society in which assigned tasks are done with faultless precision.

The pinkish horse conch on the beach this morning is fully alive, crawling along leaving faint prints in the wet sand. Not rare here, but not common either, they usually live in deep water, sometimes coming ashore. This conch, nearly five inches long, is a juvenile; adults grow to much larger sizes. A translucent pink, the shell is a spiral sculpture, a work of art. And the animal inside, moving seaward now, is our largest snail.

At sunset, intensity and color of light change frequently as the sun sinks into the western horizon. Sun rays strike objects differently and fleetingly. For a few moments, boats in the small craft harbor were richly bathed in reddish light, hulls, masts and glass reflecting the sinking sun. Now below the horizon, the sun fills the sky with lustrous color, inflaming gray clouds, soon to fade into spectral night.

Under ashen gray skies this morning, shorebirds roost on the beach, immobile, sodden. Gulls, terns, skimmers, plovers, willets—all wait in the warm rain. Each bird faces into the wind, as they always do even in the mildest breeze, prepared for sudden flight from danger. When in jeopardy, they elevate into the wind, veering left or right, quickly distancing the threat. Instinctive, preparing to flee from danger has been learned through the ages.

As beach walkers approach, it is the black skimmer which sounds the first alarm, warning other shorebirds in the flock of approaching danger. Barking low-pitched resonant notes, they fly away. Also fearful of humans, great blue herons flee early, croaking "fraak, fraak." Laughing gulls, more calm at a walker's intrusion, walk away, with single-note complaints of "kuu." Remaining with the gulls, royal terns nervously call a shrill "key, key."

$\mathcal{S}$everal common pigeons are on the beach today, eating sand needed for their digestive tracts. They are merely "city pigeons" like the ones occupying nests in buildings, distant cousins of those pesky ones infesting the Piazza San Marco in Venice. First domesticated 6,500 years ago, they were once raised for their meat. Although a beautiful iridescent bronze-purple-green, pigeons are now rebuffed by many as civic nuisances.

Their disappearance and reappearance is magical! A few minutes ago sanderlings were feeding peacefully at water's edge. Alarmed by a walker, the flock quickly flew seaward. For a moment, they were totally visible, gray bodies, striped wings. Then veering eastward, they vanished. Pivoting seconds later, their wings flashed in the sunlight. Next, flying to the east, the sanderlings once again simply dissolved, camouflaged in nature's alchemy!

There was a time when a shorebird's death brought deep sorrow. Lying lifeless at water's edge today, a royal tern retains its majesty. There is sense of loss, of course, but also insight to a commonplace event. Creatures are born, grow, and die. They eat and are eaten. In nature and in life, change is normal. Not in charge, we merely witness the beach's remarkable economy, which celebrates another day with renewed life.

The Beach in

Summer

At water's edge today lies a robust jellyfish, undamaged but stranded by the receding tide. Pulsing, throbbing with energy, one notices without a magnifying glass, its vitality is vibrant. Surely it will live and escape into the next tide. Nearby in black silt, scores of almost microscopic insects hop. Another life form exists beyond our knowing! It is the poet Blake who instructs us, ". . . Everything that lives is holy; life delights in life."

America: A Prophecy by William Blake.

*I*t is not easy to leave the beach this fine June day. How pleasant it would be to face the mild breeze, to digest the beauty of whitecaps on a cobalt sea, to do nothing but sit and watch. But as Thoreau wrote when he left Walden Pond, "I had several more lives to live." He could spare no more time for that adventure. For the moment duty calls, but there is solace in knowing that beach and sea promise delightful tomorrows.

Walden by Henry David Thoreau.

At high tide today two horseshoe crabs are locked in an ageless mating embrace. Burrowing in wet sand, the female buries about 200 eggs, fertilized by the male, positioned on her back. In some 20 minutes, the act is finished with the female returning to deep water, still locked to the male. The scene witnessed today has been repeated countlessly for 200 million years even prior to the appearance of mammals and birds on earth.

Today the sun rises at 6:26 a.m. and will set at 7:30 p.m. High tide is 10:51 a.m., and low tide at 10:02 p.m. This is the arithmetic of today—those diurnal changes that have occurred 27,740 times during my life and have occurred countless trillions of days since creation. The relevance of this day is that we experience it now, feeling the warmth of the sun, seeing the beauty of the sea, celebrating its gladness at this very moment.

At high tides during August and September, stingrays swim within a few feet of the shoreline to lie motionless on the sandy bottom for a couple of hours. Fewer are seen this year than last. Since they are live bearers, giving birth to about six little rays a season, are these females that come in for the event? Are sand flats less risky for small rays than predator-filled deeper water? As the tide ebbs, the rays flow away with it.

Great blue herons possess a certain regal dignity. Today two in ankle-deep water stand erect, motionless, awaiting small fish to swim by. Herons tolerate people, but don't like them to be close. When a walker nears, the herons fly away on huge wings, loudly croaking in raucous, hoarse complaints about the disturbance. Later they are seen slowly walking in the shallows, their 4-foot height distinguishing them from several little blue herons nearby.

The imprint of God is seen in all His creation—the grandeur of Earth, the immensity of space, the intricacy of an atom. We find it here on the beach, when we take time to look. Pick up a shell, there He is! His grandeur is present, not only in the regal heights of the universe, but also in commonplace, everyday things.

"Nothing will compare with the early breaking of day upon the wide ocean," wrote Richard Henry Dana in Two Years Before The Mast. Confirming Dana, a British sea captain who had observed years of daybreak said to me this week, "Every sunrise at sea is different." So it is ashore too. This morning gray streaks appear, then pale pink clouds, a rosy horizon, next a red ball of a sun, and finally orange-hued sky.

*T*he west wind snaps lines against metal masts of two catamarans, at rest on the beach, making euphoric sounds. Indeed, a company markets the sounds on a tape designed to calm listeners. Today the notes are that of wind chimes in a gentle breeze, but in the small craft harbor where there are many sailboats, sometimes they are a harsh chorus. Was it T. S. Eliot who wrote about a "perpetual Angelus" bell? That is the sound heard today.

"The Dry Salvages," *T. S. Eliot: The Complete Poems and Plays,* Harcourt, Brace & World, Inc., New York, 1971.

Gossamer webs, spread beneath the stern canopy of a friend's boat, are delicately beautiful traps, spun by spiders. The silky, sheer strands form intricate geometric patterns; others perhaps delighted Euclid in his time. What engineers these spiders are! Will the web trap an insect? We do know that the web is strong, resilient enough to withstand wind and rain, and will stay in place for days.

The question is often asked, "Why do mullet jump?" And the answer, according to Dr. Bob Shipp, is ". . . nobody really knows." It is thought that certainly mullet jump out of fear when threatened by a predator or if surprised by a sandbar. But isn't it possible that mullet may jump simply for the joy of flight? Later this summer even when mullet swim in larger schools, often one will jump. And for what reason?

Guide to Fishes of The Gulf of Mexico by Dr. Bob Shipp, University of South Alabama, 1986.

Memory accompanies us along strange paths. "We cannot forget anything," John Muir wrote. "Memories . . . may sleep a long time, but when stirred . . . they flash into full stature." *So it is with today's walk, calling to mind another beach half a world away, submerged in the brain, but not erased by all the years. A forgotten tropical beach, stunningly beautiful, is recalled now in detail, in Hamlet's words,* "from the table of my memory."

A Thousand Mile Walk to the Gulf by John Muir (written 1867, published 1916). Hamlet by William Shakespeare.

At water's edge a little fiddler crab, only an inch wide, is troubled by the gentle wavelets rolling ashore. Each wavelet tosses it upside down on the sand beach. Righting itself, the crab crawls seaward once again. In the wake of a retreating wave, it leaves tracks in the wet sand which are erased at once by the next wave. Now, crawling sideways, one large claw lifted defiantly, it finally disappears into the safety of deeper water.

*O*bserving a moon such as the one hanging tonight above the Gulf, Plutarch believed that the dark spots were seas and the light ones were land. Lucian thought moonscapes were cheese-like. And Kepler, although harboring doubts, finally agreed the dark areas were indeed seas. Standing on the dark shore tonight, observing a magical moon, one sees its distinct oceans, knowing all the while that they are bone-dry valleys.

Plutarch, Greek philosopher, circa 100 A.D.; Lucian, Greek satirist, circa 200 A.D.; Kepler, German scientist 1571-1630.

*T*here are no waves today, only infrequent swells, which end high on the beach, never becoming whitecaps. "Slow, heaving swells" Melville once called them, noting that "these are the times of dreamy quietude." The interval between the swells this morning is almost a minute, a deep, relaxed breathing of the sea, not the mighty swells of the Pacific Ocean where even veteran sailors get seasick from the aberrant motion.

Moby Dick by Herman Melville.

Sometimes stormy weather brings magnificent frigate birds to the mainland. Today's squall has done just that! A flight of six flies west above the beach highway. Normally they remain on offshore islands living in big colonies. Easily identified, they have deeply forked tails, hooked bills, and immense wingspans. Supremely graceful, named for frigates—light, speedy sailing warships—they are called man-o'-war birds by some.

Grass grows in an alien spot at water's edge. A seed, blown here from panic grass planted near the seawall, warmed by the sun, drenched by daily saltwater tides, now sprouts a single green blade in barren sand. A spark of life survives against all odds! An 18th-century hymnist wrote, the seed is fed "by God's almighty hand; He sends . . . the warmth to swell the grain, the breezes and the sunshine, and soft refreshing rain."

"Hymn 138" by Matthias Claudius, *The Hymnal 1940*, The Church Pension Fund, New York, 1943.

A lone snowy egret wades in the sea a few feet from shore, its dazzling whiteness contrasting with green water, whiter by far than the beach's sand. Handsomest of all shore birds, the egret walks in shallow water, showing its golden slippers. A single, unblemished plume washes ashore, floating high on the wavelets. A hundred years ago, similar feathers, sought to adorn women's hats, caused the slaughter of egrets by the thousands.

*O*n the fourth day of Creation, God said, "Let the waters teem with living creatures and let birds fly above the earth . . ." He blessed them, saying "be fruitful, multiply and fill . . . the seas and let the birds multiply . . . , and so it was." The sea brims with life this morning, laden with minnows, mullet, tiny blue crabs, and more; overhead fly scores of birds. God saw that his handiwork was good. And so it is.

Genesis 1:14-22.

Just at sunset dark clouds bank in the southeast sky. On humid days such clouds sometimes appear in the north, bringing welcomed relief from the heat. But these are forbidding, more than a summer squall. The clouds mount high in the sky, overhead now, delivering first light rain, and now a deluge, wind, vivid lightning flashes, and claps of crashing thunder. Nature's fireworks brighten the night for a while, leaving a windless calm.

Two finely detailed ruddy turnstones, first seen yesterday, are busily feeding at water's edge. Easy to identify, the birds have bright orange legs, white undersides, brown markings, and black bibs—telltale signs. Sometimes called "bishop birds," because of their round collars and "eucharistic vestments," the turnstones have now found an enticing meal, a shallow cache of crab eggs buried in the warm, damp sand.

The joy of seeing an osprey fills the day! The noble fish hawk, absent from these waters for years, now returns. Sighting a target from some 60 feet above the surface, the osprey plummets downward, sharp talons piercing the water first. After a mighty splash, the bird seizes a large fish, carrying it aloft, powerful wings flapping. Flying into the oaks bordering the Sound, the osprey will devour the catch. Is its nest nearby?

Sea turtle eggs hatch on a nearby beach, attracting a crowd of onlookers. After hatching, the small turtles scurry toward the water. Normally, shorebirds attack in a feeding frenzy, devouring many of the creatures before they reach the sea. But today, because of the spectators, birds are kept at bay. Nearly all the turtles reach the water safely only to meet different perils, large fish and other predators.

The small, round rock, tossed by incoming waves, rolls high onto the beach in the morning's high tide. Now smooth from tumbling against abrasive sand as are many such stones, this pock-marked one somewhat resembles a moonscape with shaded craters, mountains, valleys, and "seas." Except for its small size, the rock might well be a planet in the sun's orbit. Instead it is but a beach pebble.

The thunderstorm which came ashore a bit after noon today brought with it lightning bolts, squally winds, and an hour of heavy rain. But it also brought a respite from searing temperatures. The haze, normal during mid-summer days, has been washed away, leaving the air clear of pollutants. The clarity brings the barrier island into sharp focus, defining individual pines on its eastern beaches.

The placid calm, which extends across the sea southward to the sea buoys this bright summer morning, is devoid of ripples. A dead calm, sailors call it. The calm's timeless, still quality will last until noon perhaps, when an offshore breeze will bring wavelets first, then full-blown waves. Henry Vaughan captured the scene in verse: "I saw Eternity . . . a great ring of pure and endless light, all calm as it was bright . . ."

The World by Henry Vaughn.

Since sunrise, two shrimp boats have been at anchor a few hundred yards offshore. Now the vessels raft together, the crew of the windward boat joining the other shrimpers, one guesses, for breakfast. After trawling all night, culling and icing the catch, the crews sleep on deck in the shade of a tarpaulin. After sunset, the boats weigh anchors for another exhausting night of shrimping on a moonlit sea.

Swimming just offshore this morning is a pod of dolphins, leisurely gliding through the sea. They are so close that their large heads are clearly visible when they surface. Of all mammals, their brains have a high ratio to body sizes. But this does not mean that they are necessarily smarter than other primates. They simply have a greater capacity to think. Diving and surfacing now, the dolphins enjoy the gift of life.

A few moments ago some fifty yards offshore, a brown pelican dived headlong into the sea in pursuit of a fish. Floating, the bird raises its bill, swallowing the fish. To pursue other game, the pelican, its broad wings flapping awkwardly, lifts itself in flight, pounding the surface with webbed feet to gain speed. Now airborne, the large bird wings eastward, wheeling gracefully in the mid-morning wind.

Writing in 1873, Alexandre Dumas mused that if all the eggs hatched in the sea grew to maturity, "you could cross the Atlantic dryshod on the backs of cod." And he might well be right! But although one codfish may produce 9 million eggs in a season, only a very few survive. Such is the law of the sea. Consider the horseshoe crab, laying thousands of eggs in the warm sand of today's high tide. How few will mature!

Cod by Mark Kurlansky, Vintage Press, London, 1998.

The little pufferfish floating at water's edge this morning is now several times its original size. When frightened, it inflates itself, projecting sharp porcupine-like spines all over its body to discourage predators. A great delicacy in the Orient, the fish is also toxic and must be carefully cleaned. One guesses this puffer to be a striped burrfish, a slow swimmer armed with pointed spines, sharp teeth, and a small dorsal fin.

How fortunate to have a few plovers left here this summer, since most have flown to tundras in the far north country. Some think that Aristotle first named the bird, which alone distinguishes it. The plover at hand feeds today in wet sand, finding tiny shellfish, worms, crab eggs, and other morsels. When it is alarmed, the birds' haunting cry, both melodic and plaintive "tea-ooo-eee, tea-ahh-eee, toooo-ahhh-leeee," is heard.

Today is the summer solstice, celebrated for eons by ancient peoples, but not of much importance to moderns. Now we are divorced from changes in the sun, moon, stars, the earth. The ancients extolled the solstice, a time of fertility, marriage, and growing crops. More attuned to nature than we, beach creatures may sense change today, a time when the sun briefly stands still, then each new day becomes shorter.

*O*ne day in the 1650s, Izaak Walton wrote, "I have laid aside business and gone a-fishing." So too have some 33 men, women and children who are fishing today on Moses Pier. Despite the storm-ruined pier end, and notwithstanding the fact that no fish are caught, there is happiness and peace among the fishermen. "God never did make a more calm, quiet, innocent recreation than angling . . . ," Walton observed.

The Compleat Angler by Izaak Walton, 1593-1683.

At water's edge in a calm sea, a large school of tiny minnows, nearly a thousand, swims within easy view. Led by one minnow, all the others follow in a perfect ellipse, copying each of the leader's movements. A small shell tossed into the school causes the minnows, not to scatter as one might guess, but to gather into a denser pattern. Shared by humans, too, the herd instinct brings the fish closer together for protection.

𝒜t a distance, the butterfly appears to be a common monarch, but the bright orange of its wingtops with white black-rimmed spots and silverish marks on the underside of the wings, show that to be incorrect. A sudden frisson! It is a gulf fritillary, an infrequent but dazzling visitor to the beach! Whether the butterfly is named for the lily, or the flower named for it, who cares? The moment is one of grace.

A flock of laughing gulls is at rest on the beach today, or should they be termed differently? Collective nouns are alluring words. There are "schools" of fish, "gaggles" of geese, a "bevy" of quail, a "pack" of grouse, a "nide" of pheasants, a "muster" of peacocks, a "pod" of dolphins, a "pride" of lions, a "brood" of skimmers, and a "flight" of birds. And, with its British origin, everyone's favorite is an "exaltation" of larks.

Banked low along the southern horizon this morning are drifts of gray-blue clouds. If one did not know better, he would judge them to be mountains, much like the blue ones in Puerto Rico, Sierra de Luquillo. In clear view today are rocky peaks, valleys and passes—all only nebulous clouds. "Tis distance lends enchantment to the view," wrote Thomas Campbell, "and robes the mountain in its azure hue . . ."

Pleasures of Hope by Thomas Campbell, 1777-1844.

"Summer afternoon—summer afternoon; to me those have always been the two most beautiful words in the English language," said Henry James. The words do bring thoughts of carefree days, leisurely spent, sun-filled, light-hearted. Perhaps he meant one like today, with white-capped waves rolling ashore driven by a cooling breeze, children building sand castles, birds flying by, and clouds floating overhead filling the northern sky.

A Backward Glance by Henry James.

Reacting to a beach walker's approach, a stingray swims into deeper water on its graceful bird-like wings, gliding through calm, clear water. Earlier it had been lying torpid near the beach in shallow water, silhouetted against the white sandy bottom. If today's land creatures once lived in the sea, as scientists believe, in how many eons will stingrays evolve into birds, flying lithely through the air?

Least tern nests, surviving the onslaught of beach tractors last week, are this morning marked by a score of orange flags. Who placed the brightly colored, protective banners is a mystery, but the unknown benefactor has saved the remaining nests from destruction. Veering to the south today, tractor drivers avoid the flagged area. About 200 terns fly out over the Sound, unaware of the benign act that saves their nests and chicks.

This is the time of the year when least terns, seeking to protect their nests, are most aggressive in diving at walkers. Defending their turf, terns begin attacking people ("zeek, zeek, zeek") long before a trespass occurs. Individuals, innocent victims, sometimes sustain painful scalp wounds from the irate little birds. However, when under attack, walkers learn to lift a stick a few inches overhead, a shield which terns respect.

The shallow water beneath the low pier is a virtual aquarium, a stage upon which throngs of sea life appear briefly and then vanish into deeper water. Schools of ground mullet, fingerlings and larger ones; two stingrays; a few blue crabs; minnows galore; and small, unidentified fish flashing brilliantly white from time to time in the bright sun—all recoil in fright at the approach of four dolphins feeding lazily on this late summer afternoon.

*P*ainting seashells and flowers, Georgia O'Keeffe became a major American artist. Like all beachwalkers, she picked up shells, rocks, and bits of wood that struck her fancy. "I have used these things," she wrote, "to say what is to me the wideness and wonder of the world as I live in it." And use these things she did, creating on canvas, "the poetry of things"—clam shells, flowers, rocks, bird feathers, bones— all verses on canvas.

"Two Calla Lilies on Pink" by Georgia O'Keeffe.

*B*oth blind and deaf, Helen Keller once wrote, "What we have enjoyed, we can never lose." We see a sunset, a moonlit mountain, a calm or stormy ocean, holding these visions in our hearts, she said. "All that we love deeply becomes a part of us," Keller reasoned. A faint sunrise this morning may become such a memory. Shorebirds awakening, a glassy calm sea, a hushed stillness—all a serene moment at dawn of day.

The Story of My Life by Helen Keller, Dover Publications, New York, 1996.

Like the rising and setting of the sun, there is a great certainty in the flow and ebb of tides. Since the beginning of time, since Creation, oceans have risen and fallen in a timeless tempo, a cadence bringing life to the beach. "The sea is the great purifier . . ." observed John Burroughs; "What a cemetery, and yet what healing in its breath! . . . How destructive, and yet the continents are its handiwork!"

Signs and Seasons by John Burroughs, 1837-1921.

$\mathcal{A}$longside the beach road, a cluster of mushrooms (or are they toadstools?) grows in the aftermath of last week's rains. Delicately small, the buttons are pearly white, atop flimsy stems. Beneath the caps, the gills are artful ridges, best seen with a magnifying glass. Carefully excavated, the root system is a maze of thread-like strings. Whether edible or not, the finely detailed fungi are pleasing works of art.

Later tonight a new moon will rise, but now as the sky darkens thousands of stars appear. So immense is our galaxy that in addition to the sun there are a 100 billion more stars. And other galaxies, unknown now, may well exist in the infinity of space. Standing on the beach tonight, one scans the immensity of the universe. Why is the small planet Earth, among endless astral bodies, alone endowed with life?

The host of least terns on the beach today, including scores of fledglings, marks the near end of a successful nesting season. These are the birds which have survived countless threats through spring and summer from both humans and roaring beach tractors. Now, a month or so away from migration, the young birds daily gain strength and experience for arduous flights to faraway winter beaches in more southerly climes.

Consider the comely whelk shell, perfectly proportioned, protecting the frail snail that it houses. At the beginning the whelk is very small, but as it grows the shell increases in size with the body whorl and aperture becoming larger while the apex and shoulder remain small. With continued growth additional material appears around the opening. The result is the graceful shell held in hand today, an evolving miracle.

*B*asking on the seawall in bright afternoon sunlight is a slender green lizard perhaps four inches long. Upon the walker's approach, it disappears into a concrete crevice only to reappear moments later, quizzically watching the intruder. Its green color matches verdant weeds nearby. A reptile (sometimes called a "skink"), the lizard looks prehistoric. Indeed, its forebears were on Earth long ago in the Triassic Period.

Summer brings a fullness of time to the sea, a fruition of all seasons, a time of teeming life. Wildflowers bloom in profusion. Mating and nesting are past. Fledglings leave nests to learn to fly. Mullet and all other fish abound. Shrimp mature. Countless minnows swim by. Life is abundant everywhere. An erupting cornucopia, the sea becomes mother of life in summer, pulsating with vitality.

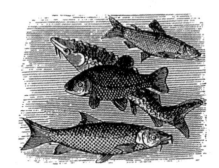

*B*yron writes of "eternal summer," but the days of bright summer on our beach are numbered. The season's end nears, and it is evident in the apperance of a few migrating birds, early perhaps this year. Some may wish to preserve the beauty of this day, staying season's change, retaining summertime forever. Oliver Wendell Holmes has a solution, "For him in vain the envious seasons roll, who bears eternal summer in his soul."

Don Juan by Lord George Gordon Byron, *The New Oxford Book of English Verse*, Oxford, 1984.
"The Old Player" by Oliver Wendell Holmes.

QUAIL RIDGE PRESS
P. O. Box 123 • Brandon, Mississippi 39043
Phone 1-601-825-2063 or 1-800-343-1583
E-mail: info@quailridge.com • Website: www.quailridge.com